FLORA OF TROPICAL EAST AFRICA

BERBERIDACEAE

R. M. Polhill

Shrubs or herbs, the latter often with tubers or rhizomes. Leaves alternate or radical, simple or compound; stipules absent. Flowers variously arranged in panicles, racemes, clusters or solitary, regular, ☿. Sepals and petals usually similar, in 2–several series, hypogynous, free, imbricate or the outer valvate, rarely absent. Stamens 4–9, opposite the petals, hypogynous, free; anthers 2-thecous, opening by longitudinal slits or valves. Ovary 1-locular; ovules basal or ventral, anatropous, few or sometimes numerous. Style short or absent. Fruit a berry, achene or capsule. Seeds with copious endosperm; embryo small or long.

BERBERIS

L., Sp. Pl.: 330 (1753) & Gen. Pl., ed. 5: 153 (1754)

Evergreen or deciduous shrubs or subshrubs. Branches usually with simple or compound spines, which subtend short shoots bearing the leaves and flowers. Leaves in fascicles or whorls, sessile or shortly petiolate, simple or with 2 vestigial leaflets at the base; blade toothed or entire. Flowers solitary or in fascicles, umbels, racemes or panicles. Sepals in (1–)2–3(–4) series or whorls, petaloid, usually with 1–2(–3) smaller bracteoles below. Petals (5–)6, shorter or longer than the sepals, with 2 glands near the base. Stamens 6; filaments ± dentate at the apex, the connectives sometimes produced; anthers opening by valves. Ovary with 1–12(–15) basal ascending ovules. Fruit a berry.

About 450 species, mostly in the northern hemisphere of the Old World with a few species extending into Malesia, also North and South America. One species in tropical Africa.

B. holstii *Engl.*, P.O.A. C: 181 (1895); Sprague in Hook., Ic. Pl. 31, t. 3021 (1915); T.S.K.: 9 (1936); T.T.C.L.: 70 (1949); Milne-Redh. in Mem. N. Y. Bot. Gard. 8: 215 (1953); Wild in F.Z. 1: 171, t. 24 (1960); Ahrendt in J.L.S. 57: 102 (1961); K.T.S.: 58 (1961). Type: Tanganyika, Usambara Mts., *Holst* 427 (B, holo.†)

Glabrous shrub, 1–3 m. tall. Young branches partially tinged dull red-brown, with many short shoots, 3–10(–15) mm. long, subtended by 3–5-fid spines; spines 1–4 cm. long, sulcate below. Leaves crowded, shortly petiolate, 3-foliolate; lateral leaflets very reduced, subulate or filiform, 1–3(–4) mm. long, persistent; terminal leaflet oblanceolate to obovate, 2·3–7 cm. long, 0·9–3·8 cm. wide, mucronate, entire or usually with several to 20 spiny teeth, coriaceous; venation open reticulate, with the lateral veins very oblique at the base, more spreading above; petiole 1–2 mm. long. Inflorescences of 8–15(–24) flowers in little-branched irregular panicles,

2·5–7·5 cm. long; bracts mostly narrowly triangular, up to 5 mm. long, with the lowermost ones often large and leafy. Flowers yellow, sometimes tinged red; perianth-segments in 5 series of 3; inner sepals the largest, spreading, rounded at the apex, 6–7 mm. long, 4–4·5 mm. wide; petals 6, brighter yellow, obovate, 4·5–6 mm. long, 2–4·5 mm. wide, with 2 glands near the base. Ovary narrowly ellipsoid, with 2–4–6 ovules. Berry ellipsoid, 8–12 mm. long, 6–7 mm. across, plum-red to dark purple, pruinose, perfecting 1–4 seeds. Stigma persistent. Seeds 5–7 mm. long, 2–3 mm. wide, brown, finely rugulose. Fig. 1.

UGANDA. Karamoja District: Mt. Moruongole, 11 Nov. 1939, *A. S. Thomas* 3284! & Mt. Moroto, Sept. 1956, *Wilson* 268! & June 1963, *Tweedie* 2660!

KENYA. Northern Frontier Province: Ndoto Mts., Sirwan, 1 Jan. 1959, *Newbould* 3401!; Mt. Kenya, June 1909, *Battiscombe* 92!; Masai District: S. slope of Mau escarpment, 3 Sept. 1936, *D. C. Edwards* 63!

TANGANYIKA. Mbulu District: Mt. Hanang, Gendabe, 11 Feb. 1946, *Greenway* 7693!; Morogoro District: Uluguru Mts., Lukwangule plateau, 13 Mar. 1953, *Drummond & Hemsley* 1553!; Iringa District: 21 km. S. of Dabaga, Idewe Forest Reserve, 20 Feb. 1962, *Polhill & Paulo* 1546!

DISTR. **U**1; **K**1–4, 6; **T**2, 3, 6, 7; highland areas from Somali Republic (N.) to Malawi and Zambia (Nyika Plateau)

HAB. Open upland woodland, edges and glades of upland rain-forest, upland evergreen bushland; 1500–3450 m.

SYN. [*B. tinctoria* sensu A. Rich., Tent. Fl. Abyss. 1: 10 (1847), *non* Lesch.]
[*B. aristata* sensu Oliv. in F.T.A. 1: 51 (1868), *non* DC.]
B. aristata DC. var. *subintegra* Engl. in E.J. 28: 389 (1900). Type: Tanganyika, Uluguru Mts., *Goetze* 280 (B, holo.†)
B. petitiana C. K. Schn. in Bull. Herb. Boiss., sér. 2, 5: 455 (1905); R. E. Fries in N.B.G.B. 9: 319 (1925); Ahrendt in J.L.S. 57: 104 (1961). Type: Ethiopia, Menisa, *Quartin Dillon & Petit* (W, holo.!)
B. grantii Ahrendt in J.L.S. 57: 103 (1961). Type: Tanganyika, W. Usambara Mts., *D. K. S. Grant* 19 (K, holo.!)

NOTE. Closely related to the Himalayan *B. aristata* DC., differing in the characters noted by Wild in F.Z. 1: 173 (1960).

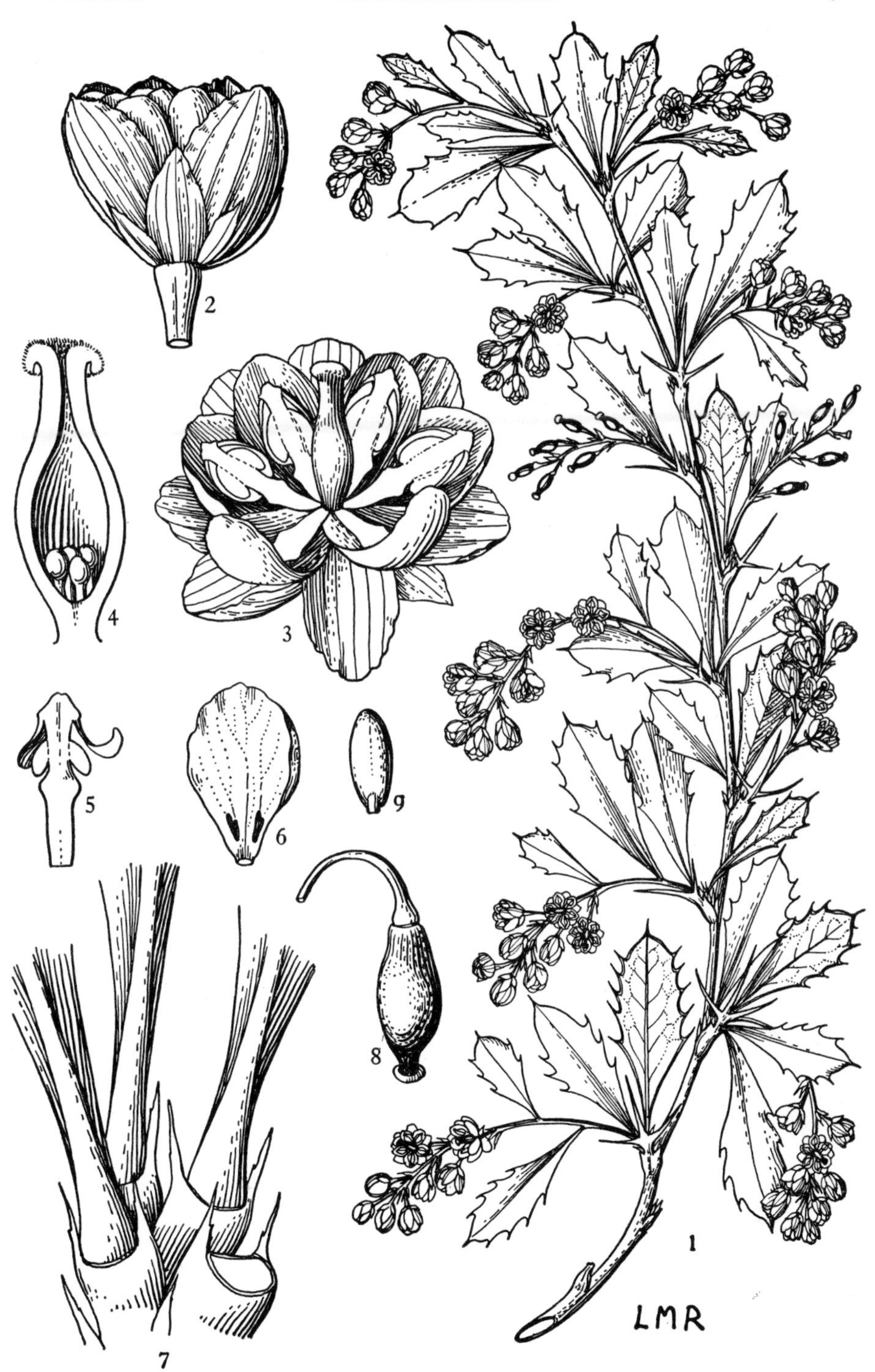

FIG. 1. *BERBERIS HOLSTII*—**1,** flowering branch, $\times \frac{2}{3}$; **2,** flower, $\times$ 4; **3,** flower expanded, $\times$ 4; **4,** section of gynoecium, $\times$ 8; **5,** stamen, $\times$ 4; **6,** petal, $\times$ 4; **7,** leaf bases, showing vestigial lateral leaflets, $\times$ 4; **8,** berry, $\times$ 2; **9,** seed, $\times$ 2. 1–6, from *C. G. Rogers* 415; 7, from *Brass* 17325; 8, 9, from *Drummond & Hemsley* 1553. Reproduced by permission of the Editors of " Flora Zambesiaca ".

INDEX TO BERBERIDACEAE